English	water	ˈwôtər
Chinese	水	shuǐ
Hindi	पानी	paanee
Spanish	agua	agua

Cartoon characters created by Navya Saini and Manish Saini. Edited by Anubha Sharma. Illustrated by Juan Carlos Santana.

English	cloud	kloud
Chinese	云	yún
Hindi	बादल	baadal
Spanish	nube	nube

First Printing, 2018
ISBN 978-1-949072-00-6
Library of Congress Control Number 2018912368

Self published
Santa Clara
CA 95050, US
https://abc4pi.com
abc4pi@gmail.com

Aata Baata Ciabatta

Water
go
Round

By
Manish Saini

Illustrated by
Juan Carlos Santana

English	ocean	ˈōSHən
Chinese	海洋	hǎiyáng
Hindi	सागर	saagar
Spanish	océano	océano

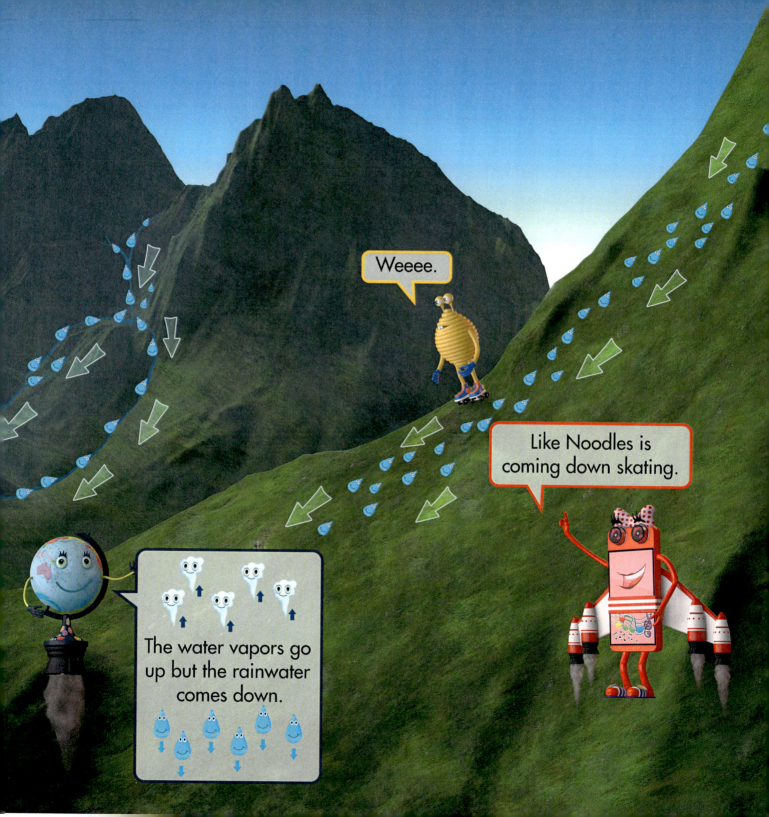

English	rainbow	ˈrānˌbō
Chinese	彩虹	cǎihóng
Hindi	इंद्रधनुष	indradhanush
Spanish	arco iris	acro iris

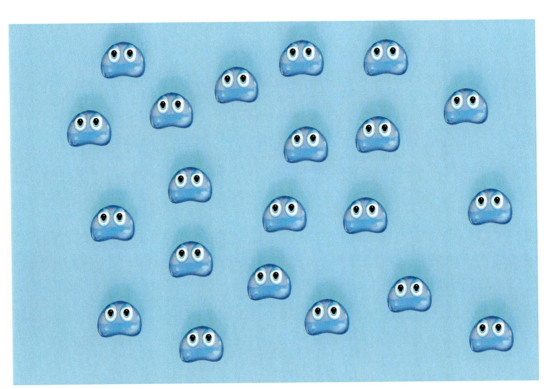

English	raindrop	ˈrānˌdräp
Chinese	雨滴	yǔdī
Hindi	बारिश की बूंद	baarish kee boond
Spanish	gota de lluvia	gota del lluvia

Do you know the shape of the raindrops?

Raindrops are not teardrop shaped.

Raindrops form in a roughly spherical shape.

As the raindrop falls, it becomes like a burger bun - bottom surface is flattened and the top surface is rounded dome-shaped.

A big raindrop (larger than 5mm) will usually break into smaller raindrops.

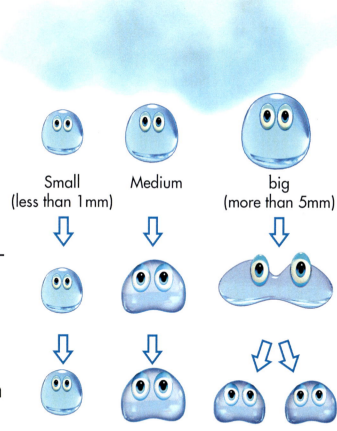

Small (less than 1mm)   Medium   big (more than 5mm)

English	sun	sən
Chinese	太阳	tàiyáng
Hindi	सूरज	sooraj
Spanish	sol	sol

Made in the USA
San Bernardino, CA
12 April 2019